For Robin & Shannon

Jenna Bryson

Text Copyright © Jenna Bryson
Illustrations Copyright © Mike Davis

ISBN 978-0-9903376-0-7

gracefromouterspace.com

GRACE FROM OUTER SPACE

Written by **Jenna Bryson**
Illustrated and Designed by **Mike Davis**

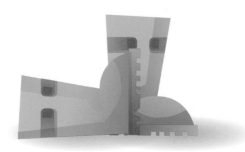

I'd like you to meet a little
girl named **Grace**.

She's just like you,
but she lives in outer space.

Yes, she lives among stars,
on this big spaceship.
And despite it s grand size,
it flies fast: **zip zip!**

Here's her little bed with a
window up high.
She loves to fall asleep
watching *stars* whiz by.

ROCKETPOD 042

DOG TREATS   SPACE SUITS

She has a rocketpod,
a spiffy **spacesuit**.
She even has a pet:
a dog named Moonfruit.

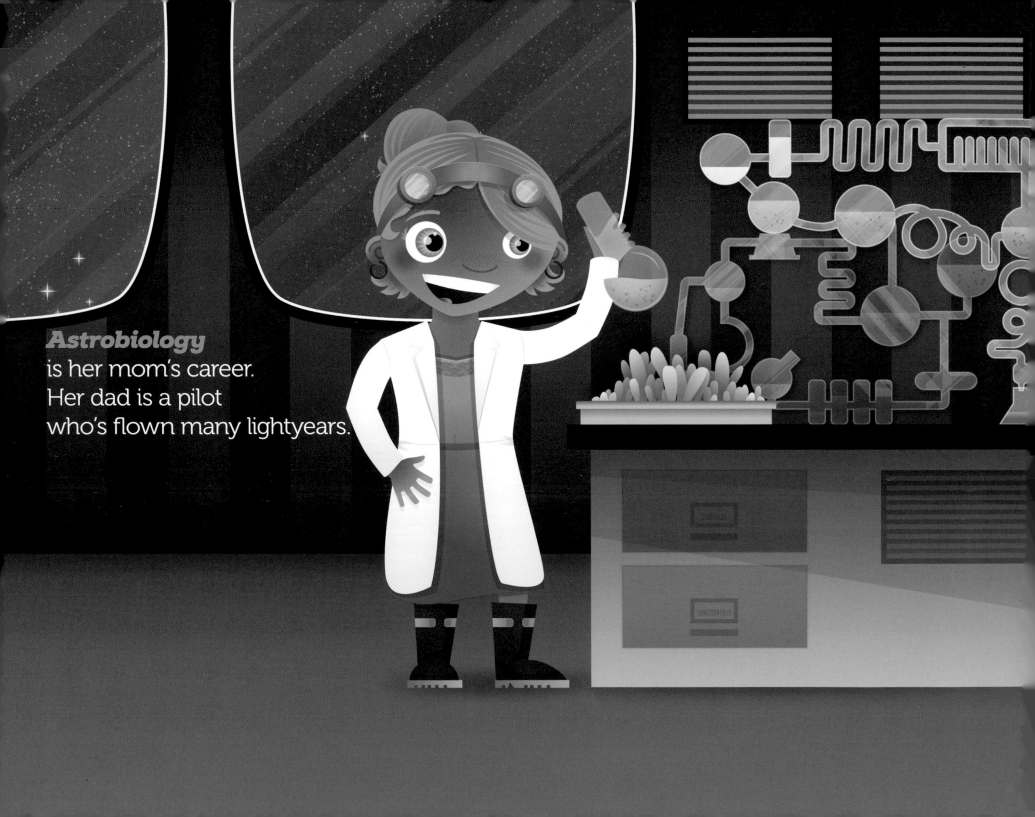

Astrobiology
is her mom's career.
Her dad is a pilot
who's flown many lightyears.

They roam the universe,
visit *galaxies*,
explore solar systems,
and cosmic mysteries.

Little Grace, she says,
"I want to touch the **sun!**"
"Oh, my stars!" she exclaims,
"That would be so fun!"

STAR:
SUN

DIAMETER:
1,390,000 km

MASS:
1989e30 kg

TEMP:
9980.33° F

But her mom says,
"You can't. That would hurt a lot!"
"Because, my dear, it's
10,000 degrees hot."

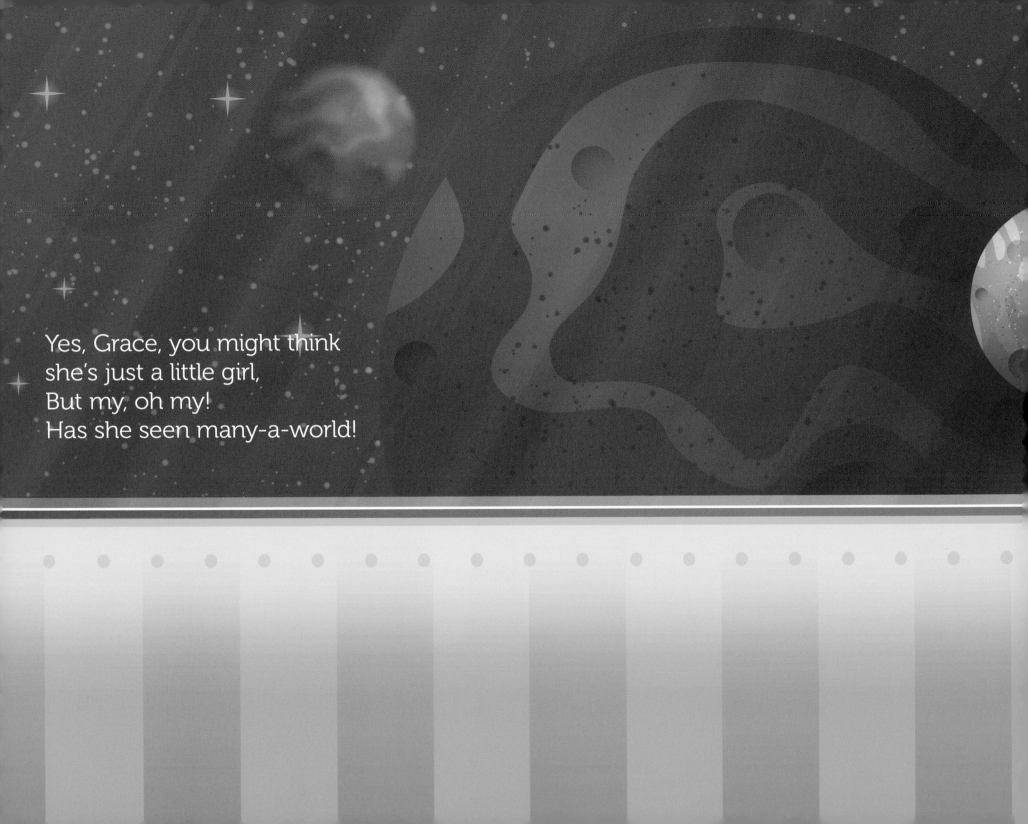

Yes, Grace, you might think
she's just a little girl,
But my, oh my!
Has she seen many-a-world!

"Oh, look! Lookie, there!
It's another Planet!
What should I call this one?
Hmm... how 'bout Janet?!"

Grace would love to be friends
with an alien.
"Maybe something strange,
a non-*mammalian*."

My alien friends

Emmy
Has hair 3.14159 feet in length

Karen
Is a fish with feet

Zoe
Is a silly Lurtle

Neil
Is a happy TigerFly

"A bird with long, long hair,
or a fish with feet.
How's 'bout a tigerfly?
Wouldn't that be neat?!"

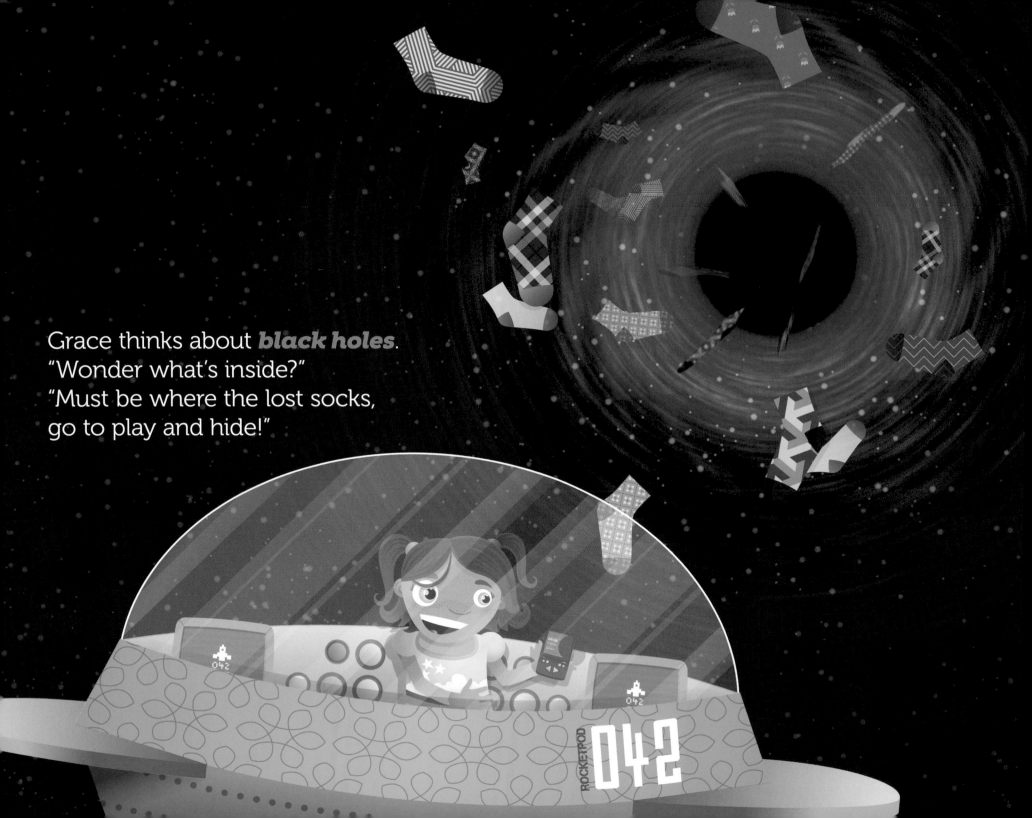

Grace thinks about **black holes**.
"Wonder what's inside?"
"Must be where the lost socks,
go to play and hide!"

She dreams about jumping, leaping, and dancing. "All around **Saturn**, from ring to icy ring!"

Grace is quite good at **science** and *mathematics*
You could say she's a real number fanatic!

AT WHAT SPEED
IS LIGHT TRAVEL?

"*The speed of light* is the best,
I must confess.
One-hundred-eighty-six-thousand
MPS!"

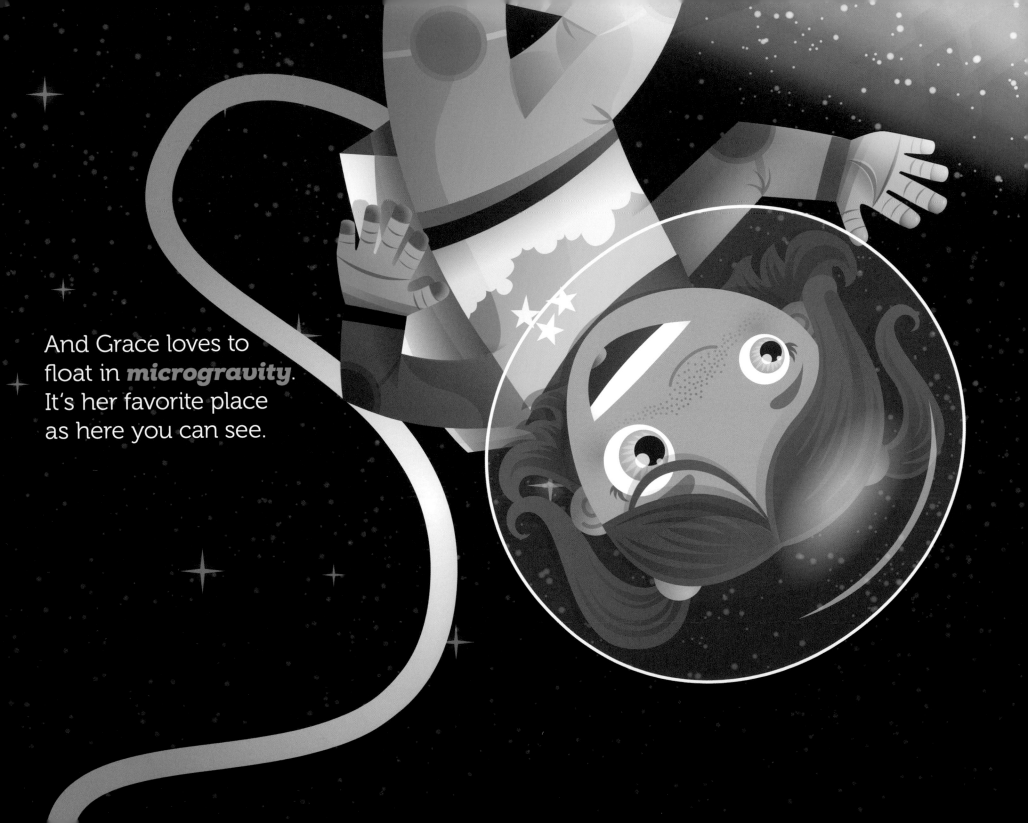

And Grace loves to
float in *microgravity*.
It's her favorite place
as here you can see.

She watches the **comets**
all racing right past.
With their vaporous tails,
they fly by so fast.

And if you are wanting
to see baby *stars*,
well, this is the kid
who knows just where they are.

"The **nebula** is
like a star nursery.
It's made up of dust, gas,
and *plasma*, you see?"

Grace has seen the **universe**
from far to near.
She says,
"Space really is the final frontier."

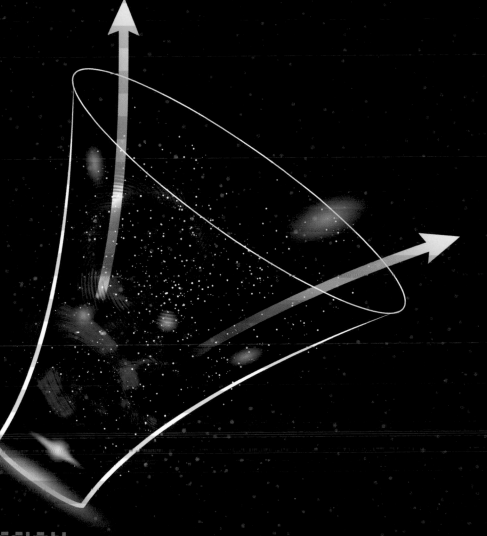

EMISSION:

DARK ENERGY

IN PHYSICAL COSMOLOGY AND ASTRONOMY, DARK ENERGY IS A HYPOTHETICAL FORM OF ENERGY THAT PERMEATES ALL OF SPACE AND TENDS TO ACCELERATE THE EXPANSION OF THE UNIVERSE.

"It might expand onward, for eternity! Thanks to some stuff that we call *dark energy*."

PLANET TRACKER 8000

"And who knows
in the future what we might find.
Maybe some others,
beings like our own kind..."

PLANET:
MERCURY

DIAMETER:
4880 km

MASS:
328.5e21 kg

TEMP:

"...they would live on a planet
just like your own
Smack-dab in the heart of
'The Goldilocks Zone'."

"If it's filled with
lots of water, and **O2**,
why, then, it could have life
just like me and you."

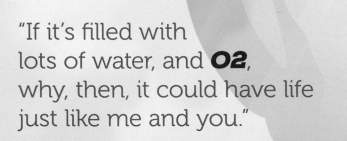

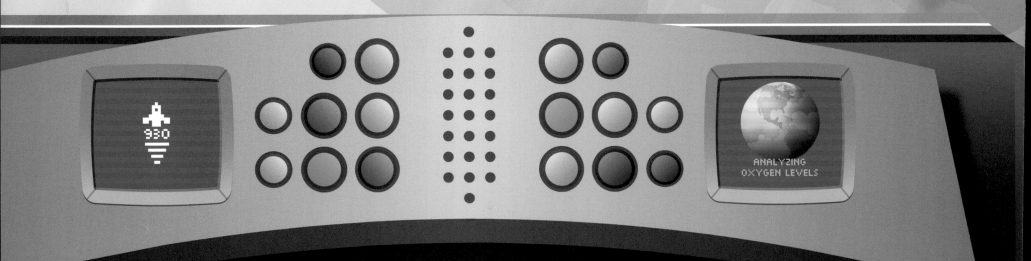

ANALYZING
OXYGEN LEVELS

"Hey, look! Lookie there!
It's a blue rock, my friend."
"I know what to call it:
how about Earth?"

The end

Glossary of Terms

Astrobiology

Astrobiology is the study of where life came from, the future of life, and where life might be elsewhere in the universe.

Black Hole

A **black hole** is a collapsed star with gravity so strong, nothing can escape, not even light.

Comet

A **comet** is made of ice and dust. Comets orbit the sun and, as they get closer to the sun, the ice melts and evaporates, giving the appearance of a "tail". Comets fly faster the closer they are to the sun, and slower the further they are from the sun. Comets in the neighborhood of Earth can travel up to 24 miles per second.

Dark Energy

Dark energy is a mysterious force that scientists think is causing the universe to expand. We can't see it, and we don't really know what it is, which is why we call it "dark" energy, but it is thought to make up about 68% of the universe.

Galaxy

A **galaxy** is a system of billions of stars, along with dust and gas, all held together by gravity. Our galaxy is called The Milky Way.

The Goldilocks Zone

"The Goldilocks Zone" is what some scientists call the habitable space around a star. If a planet is orbiting a star in just the right place (not too close to the sun, where it would be too hot, and not too far away, where it would be too cold), it is said to be in "The Goldilocks Zone". If a planet is in "The Goldilocks Zone", it would be in just the right position to have liquid water and breathable oxygen, which is what we know from our own planet that allows there to be life.

Gravity

Gravity is a mysterious force that attracts a body towards any other physical body of mass. It was discovered by Sir Isaac Newton, a mathematician physicist, almost 300 years ago. The gravity on Earth is what allows us to walk on the ground and keeps us from floating away.

Mammal

A **mammal** is any type of warmblooded animal with hair or fur that nurses its babies with milk.

Mathematics

Mathematics is the study of numbers.

Microgravity

Microgravity is when there is only a very tiny amount of gravity, such as in outer space. Astronauts on the International Space Station live and work as they float around in microgravity.

Nebula

A **nebula** is a large cloud of dust, gas particles, and plasma in space. The particles clump together, pulled toward each other by gravity, in larger and larger clumps. The bigger clumps have more gravity and attract even more dust and gas. Eventually, the pressure inside these massive clumps of dust and gas gets so high, that new stars are born.

Oxygen (O2)

O2 is the scientific abbreviation for two oxygen atoms. When two oxygen atoms combine, they become an oxygen molecule, or, O2. We need oxygen molecules to breathe, and they are also necessary for most of the life forms here on Earth.

Planet

A **planet** is a round, or nearly round, massive celestial object with its own orbit around a star.

Plasma

Plasma is what gases become at very high temperatures.

Saturn

Saturn is a planet in our solar system. It has countless icy and rocky particles, some big and some small, that orbit the planet giving it the appearance of having rings.

Science

Science is the study of the behavior and structure of the natural and physical world through observation and experimentation.

The speed of light

The speed of light is how fast light travels through space: approximately 186,000 miles per second.

Space Suit

A **space suit** is a garment worn to keep a someone alive in the harsh environment of outer space, vacuum, and temperature extremes.

Star

A **star** is a massive ball of gas that gives off heat and light. Our sun is a star.

Sun

The **sun** is the star at the center of the Solar System. It is almost perfectly spherical and consists of hot plasma interwoven with magnetic fields. The sun is a star of a type known as a G2 dwarf, a sphere of hydrogen and helium. The surface is a little under 6,000°C.

Universe

The **universe** is what we call all existing matter (everything from people to particles) and space as a whole. It's also called the cosmos. The universe is so big, that the stars in space outnumber the grains of sand on Earth. We don't know exactly how big the universe is, but we do know that it is still getting bigger.

About the Team

Jenna Bryson is the author and creator of Grace from Outer Space. She grew up in Columbia, Maryland and, although she loved "Bill Nye the Science Guy", and enjoyed "doing experiments" with her chemistry kit, science was not her strong suit. In fact, it was on-stage where she felt like she belonged.

She attended college at University of Maryland, Baltimore County, and graduated with a B.A. in Theatre with a concentration in acting. After graduation, she moved to Los Angeles, California, and quickly landed a job performing as storybook characters at children's parties. To this date, Jenna has performed at over 2,000 events throughout Southern California.

Although nothing is quite like making a child's fairytale dream come true, she continued the pursuit of her acting and music dreams, eventually teaching herself guitar and writing songs. In 2011, Jenna independently released her debut album "make/believe".

In 2012, after binge-watching anything and everything about outer space, she rediscovered her love of science and became obsessed with cosmology. Wishing that she had taken a real interest in S.T.E.M. as a student, and wondering why there were countless princesses, but zero space-faring heroines for kids, it dawned on her that she could use her songwriting and children's entertainment know-how to create a story that would not only help bridge the gender gap in children's media, but also help future generations to develop a passion for science.

She wrote the first draft of "Grace from Outer Space" in the summer of 2012. Encouraged by some friends and family, Jenna sought to publish the story as a picture book. After a few rejections and many more non-responses from literary agents, she decided she to take the book to Kickstarter. Although the first campaign in the summer of 2013 was unsuccessful, thanks to a large outpouring of support from would-be fans and media outlets (such as Mashable.com, GeekMom.com, and the NBC Los Angeles news), the second Kickstarter campaign succeeded in raising 163% of its funding goal.

These days, Jenna dreams of "stardom" not for herself, but for Grace. She hopes that this will become a series of books, a cartoon, and a live show, too (complete with original songs about science, of course). Her goal is to reach millions of little girls and boys, and ignite within them an insatiable curiosity about science, the universe, and the wonder of life.

Jenna lives in Los Angeles with her fella Jeff, their dog Gypsy, and parrot Tiki. She enjoys crafting, baking, contemporary dance, improv comedy, and reading books by Neil deGrasse Tyson (despite the fact that she often feels as if her head might explode). Links to Jenna's music releases can be found through her personal website **jennabryson.com**

Mike Davis is the illustrator for Grace from Outer Space. Born a doodler, there is no end to the number of stories that can be told of how as a child he would color on walls, paper scraps, sidewalks, or anything that was available. He can remember at a young age watching the local PBS channel and enjoying shows like "Mr. Rogers Neighborhood", where he learned about story telling, and "Mark Kistler's Imagination Station", where he got his first bits of art education.

He studied art, art history, and graphic design at Michigan State University earning a B.F.A. in Studio Art with a concentration in Graphic Design. While attending MSU, he learned a great deal of what it takes to visually communicate ideas through graphics and illustrations. His experience has led him to understand that creative design and thoughtful solutions can communicate, inform, and inspire.

He has worked as a professional designer for more than a decade in the Mid-Michigan area. While he loves designing and solving design problems his dream has always been to illustrate a children's book. So, when Jenna contacted him after seeing some of his artwork, he was so excited about the idea of working on the project. He never would have thought that posting some of his illustrations online would lead to such an amazing opportunity.

He is particularly excited about working on Grace from Outer Space because he has three daughters of his own. He hopes that through Grace, they will have a female literary heroine who will show them that it is okay to be a girl, be smart, and like science.

Mike lives in East Lansing with his wife Leanne and three daughters Emmerson, Elliot, and Laingston. He provides website, logo/branding, and many other types of creative design, as well as vibrant, beautiful, kid-friendly artwork. You can view more of his work at his official website **mdavis.in**

Acknowledgements

We would like to thank all of our backers from

KICK**STARTER**

without whom this book would not be possible. The 489 backers for Grace from Outer Space truly are a stellar bunch. We appreciate your support for this project.

We would like to give special thanks to the following backers for donating at the **Dedicated Lieutenant Commander**, **Commander**, **Captain**, **Commodore**, and **Vice Admiral** Levels.

Dedicated Lieutenant Commanders
Robert Arriaga · BNF · Dom Conlon · Sanae Richen Encarnado De Leon
Phil and Karen Drongowski · Dexter and Wanda · Christi Gell
Brock, Graeme, and Sloan McFarlane · Jennifer McKenna
Rebecca Thorpe · Ella and Emma Truong · Jaime Willis · Chris van Gorder

Commanders
Skywalking Through Neverland · The Young Family

Captains
Eva and Amelie Hart · Caroline Rediger

Commodores
Dana Hahn Samson

Vice Admirals
@datachick · Deborah Goldsmith · Zoe Tharp

Jenna would like to extend special thanks...
To our astronomy advisor **Lauren Brewer**, for not only helping us make sure that all of the science in the book was accurate and up-to-date, but for being a real-life science role model for girls, too. Congratulations on earning your master's degree!

To my parents, **Bruce** and **Fran Bryson** for their unending love and encouragement.

To **Jeff Drongowski** for shooting down all my crazy ideas... except for this one. From the very beginning, when she was nothing more than a concept and a title, you were there, and you saw what was possible; without you, who knows if 'Grace' would have ever been.

To **Gary Tharp** for his continued support and belief in me as an artist, songwriter, and now an author.

To all the grown-ups who read the book in various stages, especially **Jaime**: thank you for the insights and suggestions, all of which helped me make this book the best it could be.

To all the "Jenna Bryson music" fans who came along for the ride: it means so much to me that you chose to join me on this new and exciting journey.

And to all the little girls who let me read the story to them, especially **Brooke** and **Maggie** (the first kids to ever hear the story of "Grace" — I hope the aliens don't scare you anymore, Maggie), **Finn**, **Autumn**, **Zoe** and friends, back when there was nothing more than words on a screen: thank you for being the reason, my motivation, to make this happen. It was your giggles, your wondering eyes, your shrieks, your smiles, your furrowed brows and, best of all, your questions that pushed me to bring this book to life.

Mike would like to thank...

God for giving me my talents and abilities. For blessing me with the opportunity to use these gifts to empower young women to reach for the stars.

My wife **Leanne**, without whom I would never have considered illustrating a children's book. You are the reason that I do everything that I do. You are the person who will always answer me honestly and support me through everything. Thank you for being Supermom, CEO of everything, my design advisor, copy editor, biggest fan, and greatest supporter.

My wonderful, cheerful, inquisitive daughters, **Emmerson**, **Elliot**, and **Laingston**. Thanks for helping me come up with some great ideas for parts of the book, and for letting me read to you this great book.

My parents, **Juan** and **Martha Davis**. You've always been there for me, always pushed me to do more with my artwork. Thank you for all of your guidance, love, and support.

My friends **Jessica Fielhauer** and **Jenny McKee** for always being willing to look at, and critique my illustrations. Your excitement and support was always welcomed and such a huge help.

Thanks also to **Laura** and **Kate McKee** for indulging my crazy brain and listening to all of my crazy ideas with me, like hippo-unicorns, ostrich hockey, and any of our *"global rules."*

And to all my friends and family. Thanks for always looking at whatever crazy thing I was working on and for always listening to my ramblings.